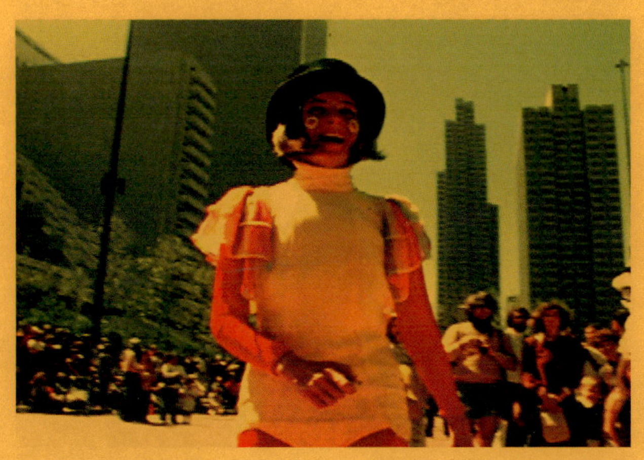

Becoming A. Banana
Written by Nika Simovich Fisher

Edited by Ian MacDougall

Film images of Banana Olympics courtesy of Steve Klocksiem

Scanned photos and editions of VILE and Banana Rag courtesy
of Anna Banana

Printed by Conveyor Studio

Typeset in Euclid Flex, Frankfurter, and Banana

Printed in June of 2023, originally written in 2022

© Copyright 2023

ISBN: 979-8-218-22127-0

BECOMING A. BANANA
NIKA SIMOVICH FISHER

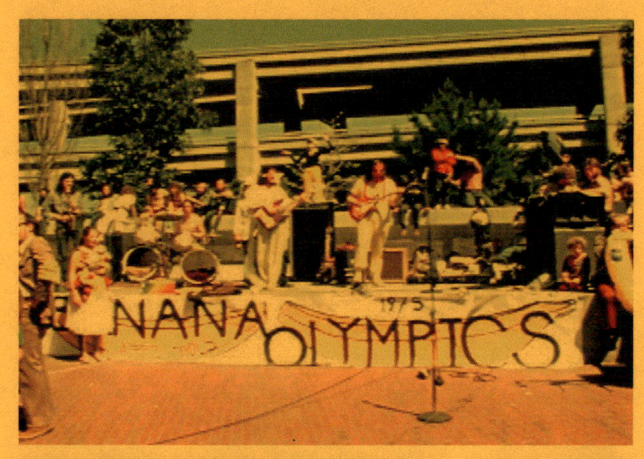

NANA OLYMPICS
1975

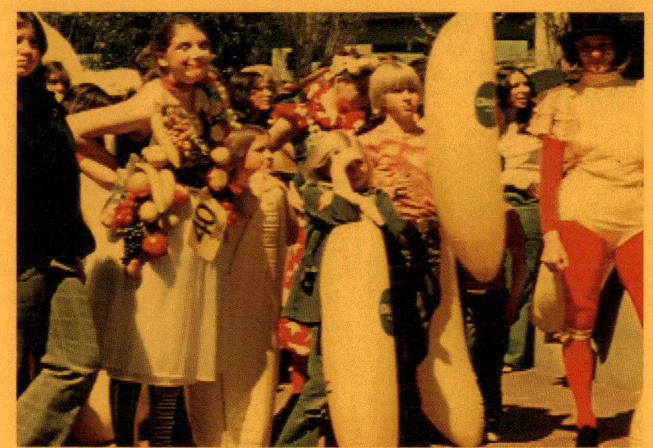

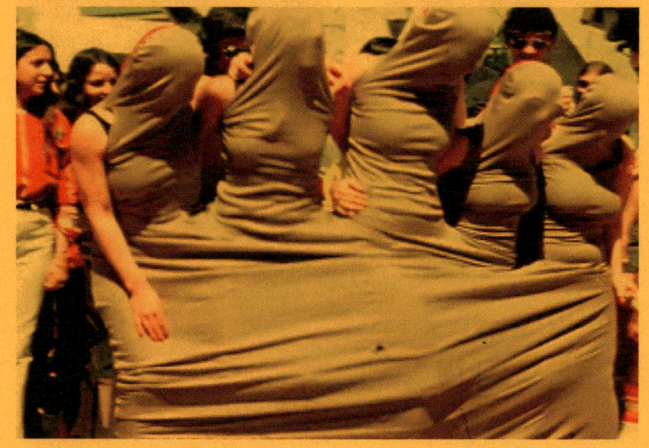

Bananas are universal. There are no banana spe-
cialists, or academic bananas, or lay bananas.
Bananas are essentially democratic. And I think to
see everything upside down is the most important
thing in art at all. If you see everything upside down,
you will see everything right.

Kenneth Coutts-Smith

BANANA OLYMPICS, APRIL 1975

The platform was draped with a white banner that read:
"BANANA OLYMPICS. APRIL FOOLS." It was situated at the end
of Embarcadero Plaza in downtown San Francisco. Long-haired
onlookers holding real bananas decorated the periphery, while
the stage was open to a revolving door of performers—poets,
dancers, comedians, and musicians. One performer sang about
bananas. The lyrics started, "Mellow, soft and yellow, and so good
for you."

Near the stage, a gathering of bell-bottomed and unshaven
participants each carried oversized inflatable bananas over
their heads. At the sound of a whistle, they tilted the balloon-like
objects first to the left, then to the right, then straight overhead
creating a banana wave.

Other characters meandered through the plaza. There was
a Mr. Peanut man with a lifelike shell. Five women dressed as a
hand—they were bound together with taupe spandex covering
their bodies and faces, forcing them to hobble awkwardly as
a human centipede of sorts. A woman with artificial fruits connected to a white dress moons the crowd, exposing striped shorts
with a smiley face made of banana appliques on her butt. Two
individuals wore fluorescent pink "DA" letters (together spelling
"Dada," a reference to the art movement) that consumed their
whole bodies save for their limbs. They spun around merrily while
spirited normies joined in, too.

The event host was the top banana. She was a mellow-yellow fruit queen roller skating through the crowd in opulent banana garb. The first layer of the costume included bright red tights and a matching long sleeve undershirt. On top, she wore a yellow turtleneck leotard with a sculptural banana hanging in the back. Ruffly rainbow sleeves accentuated her shoulders, as did yellow ribbons around her knees. She wore a black porkpie hat, and had daisies painted on her cheeks. This outfit was created with the assistance of a theater costume designer and was one of several banana outfits she owned. Her appearance showed commitment, and her name did, too. She went by Anna Banana—she had legally changed her name even—and this event was her creation.

The Banana Olympics were part of several interactive art happenings that defined Ms. Banana's practice. The event was a parody of the Olympic games, and participants were invited to join satirical track and field events, like an overhand banana throw or a parade of non-motorized vehicles. On a recent phone call, Ms. Banana, who turned 82 in February, explained that players were evaluated not on skill or dexterity, but by who did it with the most "A-p-p-e-e-l. Pun intended." Showing up as a characterized version of yourself was invited, and implementing that facet of identity was celebrated. The event was a big deal. The Bay Guardian newspaper published a full-page advertisement for it where individuals could sign up to attend, and it was livestreamed on KPFA-FM Berkeley. While turnout from the San Francisco community accounted for the bulk of the crowd, the main performers were all part of a network of avant-garde mail artists.

Mail art is a performance-based art practice experienced through the postal service. Besides that essential, postal medium, mail art embraced a mixed-media approach to art with no boundaries. Work took the form of postcards, Xerox prints, ephemeral collages, cryptic illustrations, letters, rubber stamps, postage stamps, clip art collections, newsletters, and artist zines. Many people credit its origins to Ray Johnson's New York Correspondence School and the Fluxus movements of the 1960s. These movements centered around the idea that anyone could be an artist, anything could be art, and that everyday moments of pure existence were themselves forms of artistic expression. Mail art took this idea to heart, and literally put the artwork in the hands of its participants. By sending aestheticized messages through the post, artwork became less about a finished product—a painting hanging in a gallery—and more about the exchange between two people. It democratized art—artworks in the form of mailed letters were cheap to produce—and collapsed the distinction between publishing to performance. While it progressed into a global art movement, it was originally an exchange between artists that gently poked fun at expectations about art, while simultaneously fostering meaningful relationships through art.

In many ways, the mail art community is a precursor to today's digitized social networks. These days, Meta, the company formerly known as Facebook, is swiftly rebranding the idea of the metaverse as a novel and commercialized endeavor where users can dually exist in both a digital and physical landscape of Mark Zuckerberg's creation. But the idea of developing, or reinventing, your identity through a parallel network has existed for decades.

By sending stylized messages to each other, mail artists playfully experimented with different versions of themselves, some of which even informed the identities they would adopt in their daily lives. But it wasn't just play for play's sake; over time, these different personae of the post fused with one another and with the identities becoming an amalgamation of their characters in everyday life. The mail art correspondents became the artwork themselves. Additionally, these documented exchanges toyed with the idea of temporality—mail is meant to be disposable—but the effort and process that went into sending it permanently visualizes the iterative and messy process of becoming oneself over time.

The Banana Olympics were a culminating event for Ms. Banana. "It was like I had died and gone to heaven," she said. "Here were people being creative in the same vein I was—by assuming identities and costumes that definitely got attention. We weren't so subtle."

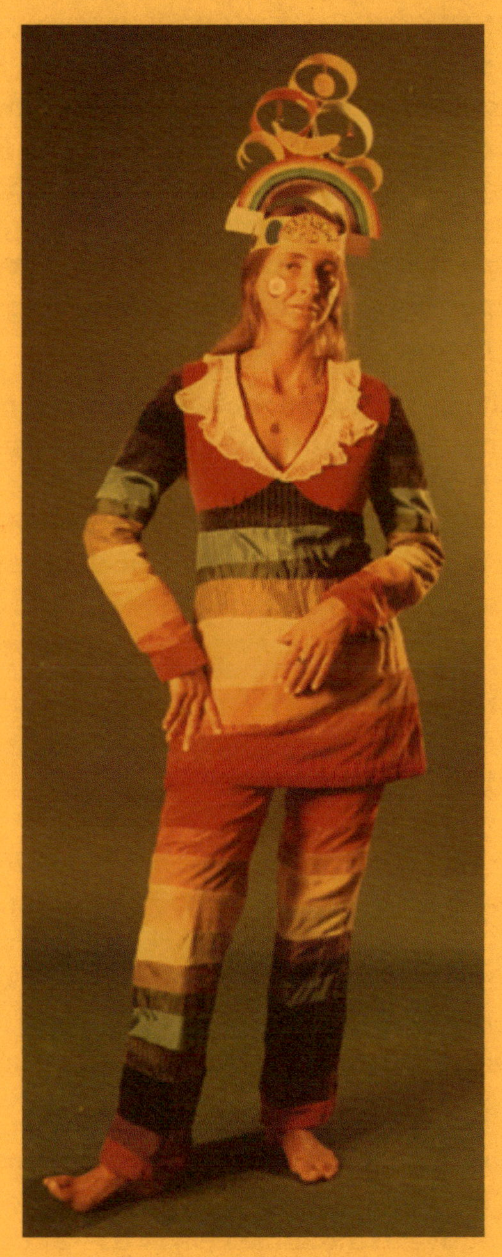

Soon after, her relationship with Bill Gaglione, her long-time collaborator and a central figure in both the Bay Area Dada and mail art scenes, would begin to rot. They were evicted from their Church Street apartment, and then she moved back to Canada. But for the moment, bananas were omnipresent.

THE TOWN FOOL QUITS, SEPTEMBER 1972

Before she was Anna Banana, Anna Lee Long lived in Canada. For ten years, she led a domestic life with a husband and daughter and family nearby and worked as a teacher in public schools. Her life then was governed by family ties, her marriage, and a series of decisions made by expectations of what a woman's life should look like. By 1969, she felt weighed down by her own ambivalence about meeting societal expectations and, like many younger people of that era, left her life behind in search of something more. "I was criticized a lot for that. But it was a survival move," Ms. Banana said. If she didn't make a change, an artistic part of her being—one that she had cherished and cultivated since youth—would have to be shut down forever, and she wasn't ready to do that.

Ms. Banana fled to the Esalen Institute in Big Sur, California. She initially attended a five-day retreat, but she came back a few months later to train as a massage therapist, supporting herself by working as the resident laundry lady. After two years, she returned to Canada to review the aftermath of her decision to leave her old life behind. Dana, her daughter, was living with her now ex-husband in a rural area outside of Duncan, where she was able to keep horses, which became a lifelong passion. Ms. Banana and her ex-husband remained on amicable terms, but he had moved on and was dating other women. Her family and friends were certain that she had "gone bananas," as she put it to me recently. In their view, which reflected the mores of the time, she must be suffering from a serious mental issue if she was unwilling to carry on happily as a wife and mother. The decision to leave her old life was difficult, but not premeditated. She didn't plan on leaving permanently initially, but during the five-day Esalen retreat, Ms. Banana had felt like she was out of her cage for the first time, and she never wanted to go back in again.

Her time in Big Sur was a moment of rebirth. When she returned to Canada, Ms. Banana began using "Anna Banana" as her moniker. The reasons for appending "banana" to her name were layered. Conveniently, it rhymed with her first name, and visually bananas were instantly recognizable and striking. She found humorous the contrast between their wholesome quality, as a nutritious and common food item, and their subliminally phallic appearance. In a 2021 interview from Fluke Fanzine she said, "It works, gets attention, has those comical overtones, and is associated with monkeys, both as animals and language."

From 1971 to 1972, she resided in a remote beach cabin outside Victoria without running water. She supported herself by giving massage workshops throughout Vancouver, Ottawa, and Toronto. Between the workshops, she focused on her art practice—first by creating batik wall hangings and later by painting geometric patterns on smooth rocks that she collected on the beach. She didn't see Dana very much during this time. The judgmental lectures and commentary from family and friends, which left her feeling ostracized, encouraged her to stay divorced from that world. While her new life left her more self-aware, she was also more isolated. Living in the wilderness, she reveled in a Thoreauvian realization—she could create a new, deliberate network through her own effort, and perhaps artwork could be the fulcrum. She began selling her painted rocks as door stops in Bastion Square, a marketplace in Victoria, British Columbia. She quickly found the commercial aspect—selling rocks as artwork—to be boring. But she found gratifying and interesting other aspects of the process: being in public, striking up casual conversations with strangers, and encouraging passersby to sign petitions against nuclear testing at Amchitka, part of the Aleutian Islands. There was more she could do, she came to realize, to draw an audience, and she soon decided to declare herself the Town Fool of Victoria.

To become the Town Fool, Ms. Banana wrote up an announcement proclaiming herself one and sent it to local newspapers and radio stations. She created a rainbow uniform to go along with it—a tiered tunic and pant combo built of alternating colors and fat fabric strips. She also wore a towering rainbow crown that said "Anna Banana" in a banana on top of a rainbow. She began wearing the getup as she roller skated through town or visited schools. Around the same time, she expanded her banana-centric identity, which started to take on the qualities it would retain during its peak, in the late Sixties and Seventies. She offered "Degrees of Bananalogy" to anyone who would share a banana joke with her and published the first edition of the "Banana Rag," a broadsheet that announced her activities, art practice, and news of her "fooling around." She would continue publishing the Banana Rag for the next four decades. Community and human connection, she was coming to appreciate, didn't have to fit comfortably within preexisting structures. She could create her own social networks, even if this one operated by way of a single node known as Anna Banana.

After sending the Banana Rag to Gary Lee-Nova, a member of Vancouver's Image Bank, he sent her a copy of The Image Bank Request List in return. The Image Bank was a seminal resource for artists in Canada at the time. Organized alphabetically by participants' names, The Image Bank was a kind of arts world classifieds section; it published requests from artists for certain items they hoped to use in their work. In one issue, the requests included "pictures of mouths," "pictures of people biting their nails, yourself biting your nails," and "pictures of politicians in action and state funerals." Although she didn't know it yet, the issue Ms. Banana received was tailor made for the artist she would become. She went on a scavenger hunt for some of the items requested in the Image Bank and sent to them what she found. She submitted a request of her own, too—for anything and everything banana-related. To her delight and amazement, she received a deluge of banana images and news clippings. Jennie Hinchcliff, a researcher, and an exhibitions and events manager at The San Francisco Center for the Book, describes the Image Bank exchanges as a pivotal moment in the history of mail art. "There was this really interesting zeitgeist going on where the search for

VOLUME NO. 11 BANANA PRODUCTIONS, 1183 CHURCH STREET, SAN FRANCISCO, CALIFORNIA 94114.

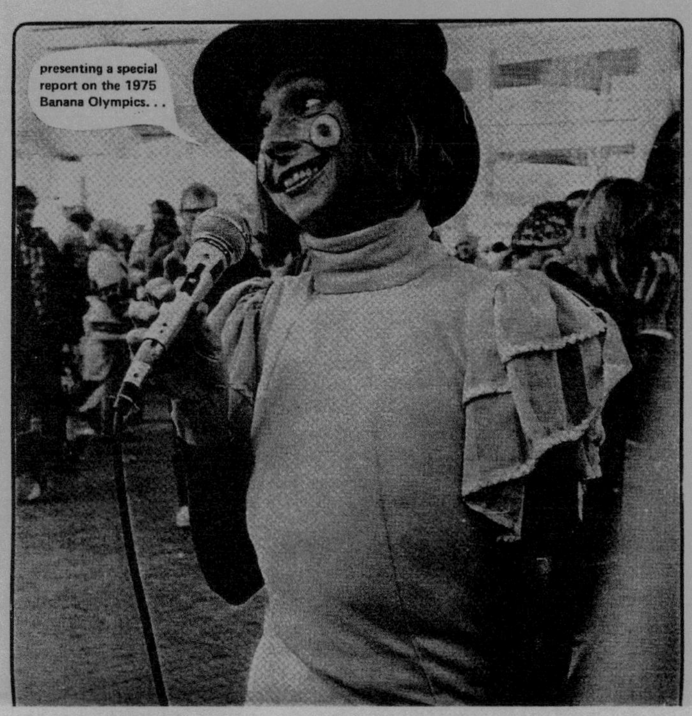

materials, and the search for a place to show your work, no matter how humble it was, gelled in the '70s and helped spin the mail art network forward," Ms. Hinchcliff told me recently. For Ms. Banana, participating in The Image Bank solidified her sense that bananas were an inviting and playful symbol, something other people could be amused by and relate to. She learned something else, too: she wasn't the only person in Canada who craved new forms of connection.

According to Ms. Banana's mail art diary, which she shared with me, her year of being The Town Fool of Victoria was an uphill battle. Everyday people on the streets of Victoria, she was coming to realize, weren't necessarily her ideal interlocutors. She grew weary of explaining the ephemerality and performative aspects of her practice to people that believed art was simply a painting on a wall.

Ms. Banana was connecting with more people through her art exchanges through the mail, and she decided to relocate somewhere more banana-friendly. The September 1972 edition of the Banana Rag announces that she's leaving Victoria. Over the next few months, she purchased a Dodge van and retooled it to double as a mobile home. Through the Image Bank network, Ms. Banana began to have regular exchanges with other artists through the mail, sending each other ideas and letters, the germ of the mail art scene Ms. Banana would soon see bloom. She mailed her correspondents her travel route and started tracking the network down to San Francisco.

SAN FRANCISCO, SEPTEMBER 1972 TO JANUARY 1981

Ms. Banana's time in San Francisco was a prolific one. She continued expanding her network and corresponding with more artists, helping one another find buyers and organizing shows for their work. She continued publishing the Banana Rag, and her Autumn 1976 issue featured illustrations created by her daughter, in addition to stamps found from a cartoon strip. She also sent through more Degrees of Bananology to anyone who contributed banana materials (photos, news stories, recipes, cartoons, songs, objects, personal anecdotes), and expanded her art practice to collaborations with the Bay Area Dadaists including a "Futurist Sound" performance. While different in media, her sent mail pieces were frequently yellow, partially typeset (Ms. Banana supported herself as a graphic designer, and became skilled at print production processes), and partially handwritten. Almost all of her mail included custom stamps; she also ran a full color stamp production service.

Through the mail art network, she met Bill Gaglione who became her partner in marriage and in art. Through their blossoming relationship, they flourished in the arts scene, independently publishing zines, organizing public art events, hosting dinner parties for the community. At one point, they even went on a European tour to share their work. They supported their practice by taking on graphic design and production jobs throughout the city.

Before he met Ms. Banana, Mr. Gaglione had helped found The Bay Area Dadaists. He went by the alias "dadaland," frequently dressed up as the letters "DA", and later adopted still another alias, "Picasso Gaglione," which he still uses today on his Meta profile. The Dadaists experimented with art pranks, including The Pink Dot Caper, in which they placed pink dot stickers around San Francisco. They sought to revive the nonsensical intellectualism of the original Dada movement, which emerged in Europe after World War I and used absurdity and humor to react to the dehumanizing force of warfare and industrialism. Marcel Duchamp's Fountain, a readymade sculpture of a urinal, is among the most iconic examples of their tongue-in-cheek philosophy.

Despite being based in San Francisco in the '70s, and possessing its own kind of politics, Mr. Gaglione's group rejected much of the overtly political aspects of the local fringe culture. One wouldn't find the Bay Area Dadaists at rallies protesting the Vietnam War with the hippies; they were as skeptical of the self-serious hippie movement as they were of the fine art world of the East Coast. "We sort of goofed on hippie culture, though our hair was long and we smoked dope too," he told the San Francisco Chronicle in 1998. He described the movement instead as a communal space for shared experimentation: "It was similar to the Internet, basically."

Ms. Banana found this ethos familiar, as it animated her mail art. The mail art network became both a refuge from the real world, but also a flight from it that developed its own internal concerns. It was also a channel of distribution for artists who didn't conform to what was fashionable among gallerists and art dealers at the time. On the East Coast, the center of gravity for the American art world, the galleries were a closed and elitist community, which made it difficult for newcomers to be seen and for different aesthetic approaches to be recognized. Out West, the cultural climate of Haight-Ashbury and The Summer of Love attracted an artistic set that, despite their eye rolling at the hippies, embraced a counter-cultural style. In mail art, they found a way to show their work on their own terms and share it with people who cared about what that work sought to achieve on its own merits, not how closely it fit the aesthetics of the important New York critics. Reflecting recently on this time period, Ms. Banana said she was focused on her artist community, doing goofy things, and enjoying herself without taking the world situation too seriously. "There is nothing you can do about it anyways," she said on the phone. The mail art network offered a parallel landscape to both the mainstream art world and current events, a space where she and her fellow correspondents could leave their mark.

Humor offered a way to deflect from the real world, but Ms. Banana and her fellow mail artists were serious about their practices. The tension between jest and sincerity highlights the way the network straddled fantasy and reality, and this philosophy ran deep in all the communications, and personal relationships that emerged from it. In 1974, after FILE Magazine, a parody of Life Magazine that focused on artwork, criticized the mail art movement as a plague on art that needed to be removed immediately due to its amateurishness, Ms. Banana started a publication of her own. VILE, itself a parody of a FILE. The magazine was intended to be a new forum for mail art. Ms. Banana sought to highlight the negative, anti-social aspects of humanity, while simultaneously documenting mail art exchanges. The first two editions ran to only 200 printed copies and were distributed through the mail art network. Subsequent editions produced 1,000 copies, which continued to be shared through the mail art network but also went to a few museums, and local university libraries.

It was through VILE that Irene Dogmatic met Ms. Banana. Ms. Dogmatic saw that mail artists were soliciting art-making materials from each other in an early edition, and perhaps unsurprisingly, she requested dog paraphernalia, though Ms. Dogmatic, whose real name is Nancy Mosen, would only later adopt her new moniker, which came about through mail art exchanges. "I've always liked dogs," she told me when we spoke earlier this year. The name highlighted her playful interest in the animal, which are a recurring theme in her paintings. Prior to joining the mail art community, Ms. Dogmatic was a self-described hippie housewife living in Oakland. By becoming Irene Dogmatic and participating in the correspondence art scene, she engaged in various happenings and art events that were entirely her own, a process of self-discovery that later prompted her, like Ms. Banana, to leave her marriage.

Pseudonyms were common among the mail art set. Like screen names or social media avatars, they at once obscured real-life identities, allowed correspondents to highlight their preferred parts of their real-life identities, and created entirely new identities. It allowed for a kind of freedom of self that some of them couldn't afford in their personal lives. "I never changed my name legally because I had to work," Ms. Dogmatic told me. "I didn't think it would be easy to drive a taxi or substitute teach as Irene Dogmatic. Or maybe I'm straighter than them." Indeed, I learned in our conversation that Ms. Dogmatic—as Ms. Mosen—had filled in as a substitute teacher in a class I took at my Bay Area high school. She had a memorable look—striking red hair worn in pigtails—but none of us, even those of us interested in art, had any idea that she had been a fixture of the local arts scene.

VILE's motto was "The worst is yet to come." The logo parodies LIFE Magazine's logo—a double parody, since FILE was itself a parody of LIFE—and the cover photo was inevitably a jarring one. (Parody is a favorite artistic tool of Ms. Banana; consider the 1975 Banana Olympics.) One cover features the artist Monty Cazazza made up to look like his heart had been pulled out. Another highlights a nude man hanging from a noose. A third highlights a woman wearing ass-less pants bending over while a clothed man gestures at her. The publication provided "Awards of Disgust" for the most grotesque pieces in each edition. In the December 1975 issue, there is an article called "Something Vile" by Richard Morris, a science fiction story about "The Eh" who transform inhabitants of Earth into "pillars of shit." There is a satirical story about how popular VILE is. "More than half of all men, women, and children over age 10 read VILE regularly or occasionally in the course of 13 issues," one statistic reads. There is a poem called "Catshit Mother Piece" by John M. Bennett, which describes a ritual of pleasuring oneself after refrigerating cat waste. (This won a VILE Award of Disgust). The main characters in many of the submissions are the people in the network. For example, "Something Vile" features Ms. Banana, Irene Dogmatic, Opal L. Nations, and other regular contributors to the mail art community. "Things were looser then in terms of sexuality," Ms. Dogmatic tells me. "It was a lot of crazy sexuality at that period of time that came out in art." The publication reads as almost a giant inside joke meant at once to titillate and repulse the outside observer.

VILE and the mail art network that birthed it embraced this tongue-in-cheek method of communicating, a mode that was at once attention-grabbing but also intentional. While the publication can be repulsive to someone not in the network, their aesthetic and tone unified them as a tribe—and suggested what qualities one needed to possess or be willing to embrace to join. Because corresponding through mail was far slower than calling someone, part of choosing to communicate this way immediately helped demonstrate the commitment of participants, and that, in turn, allowed them to develop stronger relationships with like-minded people.

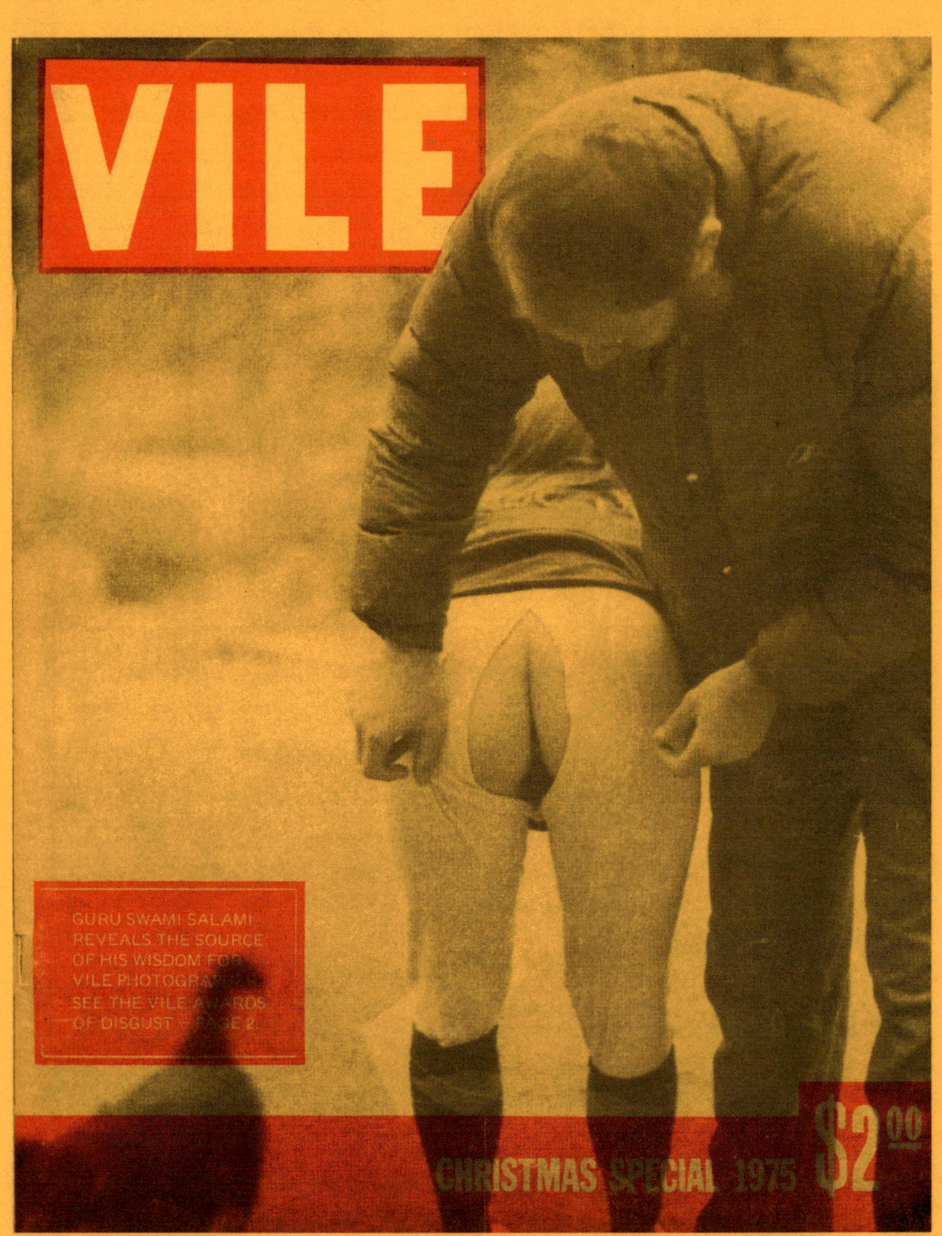
VILE
GURU SWAMI SALAMI
REVEALS THE SOURCE
OF HIS WISDOM FOR
VILE PHOTOGRAVURE
SEE THE VILE AWARDS
OF DISGUST — PAGE 2.
CHRISTMAS SPECIAL 1975
$2.00

18

Ms. Banana continued to publish VILE Magazine during her time in San Francisco. Mr. Gaglione would collaborate with her on it sometimes, serving as an assistant editor. In 1978, he convinced her to publish and finance his own edition of VILE, called "FE-MAIL-ART," which he dedicated to the work of the women involved with the mail art network, among them Ms. Banana, Yoko Ono, Cosi Fanni Tutti, Beth Anderson, Carol Berge and Alison Knowles. There were fewer women than men in the community, and despite many of the women serving as the community's motive force, the men seemed inevitably to get the greater share of attention by critics and other outsiders interested in the movement. Forty-four years later, one can still hear the incandescent irritation in Ms. Banana's voice when she describes this issue of VILE. "I was not happy because it was a completely different format," she told me. "The size, the actual size had different measurements than the originals. And then to call it 'Fe-Mail-Art'—there is nothing that says it's VILE Magazine until you get inside." She went on: "He basically got on my platform and then transformed it into his platform with his preferences, which was him." Nor was he exactly interested in women for their art. "He was a womanizer," Ms. Banana said. She recalled a party in Oakland that they both attended in the '70s. When she was ready to leave, she went to the coat closet and found "dadaland" inside making out with another woman.

Ms. Dogmatic remembers the disintegration of her friend's relationship. Ms. Banana and Mr. Gaglione continued living together, but split their apartment in half, an arrangement Ms. Dogmatic described as very "New York." "It was a messy break up," she said.

Ultimately, that arrangement was untenable, and Ms. Banana departed San Francisco in early 1981, returning to British Columbia. Yet in mail art, she had discovered that a committed network of like minded individuals could continue foster their relationships and grow as people even across distances. The mail network continued to expand; new artists would join while longtime members continued to correspond with one another. VILE continued for a time, but unlike mail correspondence, producing a printed magazine remotely was a trickier enterprise. The last issue was published in 1983, a retrospective of its history called "About VILE."

BACK TO CANADA, JANUARY 1981

In Canada, Ms. Banana continued to collect banana materials, participate in mail exchanges—with, among many others, Ms. Dogmatic—and develop, through this process, new iterations of herself. Her "Sometimes Yearly Banana Rag," which she had started publishing in June of 1972, continued long after VILE's final issue The project was more personal for Ms. Banana. Printed on 11 x 17" canary yellow newsprint—eventually she downsized to the standard, and mail-friendly, 8.5 x 11"—issues of the Sometimes Yearly Banana Rag are densely packed with personal updates on her work as an artist, and the banana content readers shared with her.

Ms. Banana continued to try to expand her network of mail artists. In the 1987 issue of the Banana Rag, she included a "Mail Art Starter Kit," in which she describes how to participate in the mail art network: "Usually a good (original work) elicits a 'good' response, but there are NO GUARANTEES in MA. You may send off something fabulous and get no response at all...you may send off some xerox copies and get a good response." Through her correspondence and the diaristic digest that was the Banana Rag, Ms. Banana continued to develop her sense of self by way of her favorite fruit. In the June 1991 issue of the Banana Rag, she highlights "the Other AB's"—other individuals and institutions who refer to themselves as a Banana. She documents the Anna Banana restaurant in Honolulu, a singer that people would sometimes send her, asking if it was her (it was a different Anna Banana), a stripper sometimes billed as the "Banana Queen of Canada," and a chef, Carmen Banana, who hosts a banana-focused cooking show. In a 2014 issue, she surfaces the artwork of Daisuke Skagami, who goes by the name "End Game" and tattoos artwork onto bananas. The issue features a photo of a banana on which he had tattooed the "Mona Lisa." She also includes a found recipe for a banana cream pie. Ms. Banana would print the forty-fifth, and final, issue of the Banana Rag in December 2015, though she would continue to produce the occasional Artistamp News, which included updates on her mail art work.

Although her time in San Francisco was relatively brief, Ms. Banana's commitment to mail art left an indelible impression on the city—not least its mail carriers. At a local zine fest in San Francisco in the early 2000s, a gray-haired gentleman approached Jennie Hinchcliff's table. In addition to working at The San Francisco Center for the Book, Ms. Hinchcliff is an active member of the mail art community who has corresponded with Ms. Banana since the 1990s. Her table highlighted a spread of mail art materials that caught the attention of the curious onlooker. "Mail, the post, that's you're thing?" he asked. "Well, I used to be Anna Banana's mail man. She would get some very interesting mail." Ms. Hinchcliff got the mailman's address, and they, too, started to correspond via post. When we spoke earlier this year, Ms. Hinchcliff reflected on the chance encounter with Ms. Banana's mailman. "The world is smaller than we think," she said. "Mail art can often remind us of that."

A NEW YEAR, FEBRUARY 2022

It's the day before Ms. Banana's 82nd birthday. She and I have been corresponding for a few months now. Her emails are always energetic and fun—she bolds and italicizes specific words, breaks sentences at dramatic moments, often repeats punctuation marks three times (???), and generously embraces ALL CAPS. It has been hard to keep up with both incoming mail, AND get on with her own initiatives, she writes, but she's doing her best as she gets older. For her BIG DAY, she has no plans, but she received two gifts in the mail. Unbelievably, they were both a PAIR OF BANANA printed SOCKS, one from her daughter, and one from a close friend. Same gesture, different socks.

Ms. Banana currently resides in Roberts Creek, a small village in British Columbia. "She's always been a little too international for where she lives in Canada," Ms. Dogmatic told me at one point. To get to her home requires a ferry from the mainland, and then a drive through the woods. When Ms. Dogmatic visited a few years back, there were five or six people living nearby—the remnants of a once thriving woodland community. Ms. Banana has been wary of adopting new technologies—she has never had a cell phone, and she doesn't have a television.

There's an irony to her generally technophobic mindset. Her mail art network in many important ways prefigures the rise of social media, albeit in a pure form, decoupled from commercial self-interest and a profit motive. The companies, like Meta, that administer social media are fond of touting how they connect the world. But Ms. Banana's experience underscores the ways in which massive, global conglomerates were never necessary to achieve the benefits that have come along with the rise of online social networks—and without the well-documented societal costs associated with the social media revolution.

Mail artists freely exchanged ideas and experimented with personas and objects that represented themselves. In her correspondence, Ms. Banana was free to embrace her most bananafied self, and the exchanges she had in this mode, she says—how she presented herself to her correspondents and what they sent in response—helped bananafy who she was outside of the mail. Consider the Banana Olympics, for example. Even after she left San Francisco, her mail art network allowed her to continue to grow as a banana, so to speak. For others, like Ms. Dogmatic, mail art was a way to realize a part of themselves they couldn't necessarily embrace as completely in their daily lives.

In this regard, mail art prefigured core elements of how online social networks function. Curator and writer Legacy Russell has extended this idea of the blurred lines between online versus offline. "IRL (in real life) falters in its skewed assumption that constructions of online identities are latent, closeted, and fantasy-oriented (e.g., not real), rather than explicit, bristling with potential, and very capable of 'living on' away from the space of cyberspace," Russell wrote in her book Glitch Feminism. "Instead, AFK (away from keyboard) as a term works toward undermining the fetishization of 'real life,' helping us to see that because real-ities in the digital are echoed offline, and vice versa, our gestures, explorations, actions online can inform and even deepen our offline, or AFK, existence."

For her part, Ms. Banana isn't thrilled to think that she played some role in the rise of social media. "The horror story of mail art is that many people have now gravitated to Facebook," she tells me on our December phone call. The effort involved in uploading a photo and some text to a website pales in comparison to the thoughtful, relatively labor-intensive gesture of writing something by hand, placing it in an envelope with a stamp, and sending it onwards. "It's too easy. And it's very superficial," she says, "I know many people are thrilled to put their stuff up on Facebook, but I find it deplorable. I hate it!"

While she is not active on social media, she keeps up with emails—one form of technology that Ms. Banana has embraced, perhaps not surprisingly, given its close kinship to physical mail. She continues participating in her postal exchanges. Over the past fifty years, she tells me, she has collected thousands of banana clippings, objects, and artworks from various exchanges through the mail. Recently, she donated most of her banana archive and costumes to the University of British Columbia Special Collections Library. She still maintains a four-drawer filing cabinet of a few left over items as part of an ongoing project she calls the Encyclopedia Bananica—her "waiting in the wings" mega proj-ect, as she puts it, where she hopes to archive the remaining four drawers of her banana-flavored collection. It will be a template, perhaps, for anybody who realizes that corporate social media, like Facebook, was never a good way to organize a social network to begin with. In the archives of A. Banana, they may discover a better, decentralized path to human communion.

A TEAM

Me & "the boys" at Intermedia Press in Vancouver, where I learne(d) the steps/procedures for print production - circa 1980, after my return from my years (? 10?) in San Francisco + 1969/70 - '80 approximately —

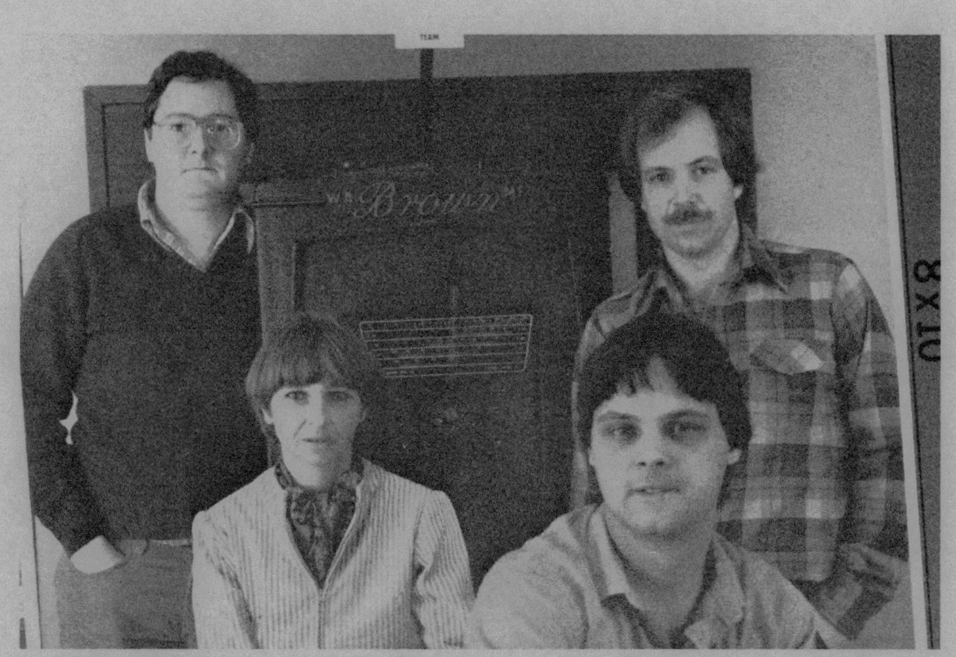

TEAM
W.R. Brown
8X10

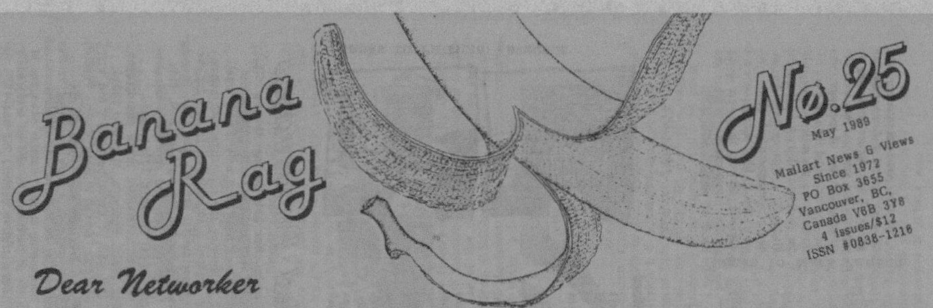

Dear Networker

According to Leavenworth Jackson, 'Time flies like an arrow. Fruit flies like bananas.' While I have experienced the truth of both, it is the former which constantly presses on me these days. A year and a half has almost passed since the death of Michael Scott. An 8-stamp Commemorative Block of stamps was issued in December, and it is time to complete the memorial book. To date only seven pages have been received for this book. For those who have been intending to do something, a reminder: 200 copies, 8½ x 11"/21.5 x 28 cm. pages plus $10 to cover binding & postage. Please let me know by the end of May if you are sending pages, otherwise I'll go ahead with those received to date.

Since there are so many topics to be covered this time, I'll sign off, with many thanks to all who have sent me wonderful banana items (see center pages), and all the correspondence and artist periodicals. It's all most amazing and very much appreciated.

Meanwhile,

Enjoy A.

Network Forum

It is with much regret that I must report the death of a most respected networker, Michael Bidner, on April 5. Concerning his mammoth Artistamp project, it is as yet unproduced. However a friend of his here, in Burnaby, Rosie Gahlinger is taking it on, and plans to complete it, sending copies to those who invested in it, etc.

Exactly when all this will come about has yet to be ascertained, as she hasn't begun to assess the material Michael prepared. In any event, it's an overwhelming project. She shipped 18 boxes from London last week when she was back there for his funeral.

Michael's box in London has been closed, and all mail concerning the project should now be addressed to Rosie Gahlinger, 5515 Jersey Ave., Burnaby, BC, V5H 2L3. A word of caution before you start bombarding her with your enquiries or suggestions . . . give her a few months to get a grasp on it.

Rosie told me on the phone that Michael had received stamps from some 3000 artists in the 10 years that he collected data for the project. In 10 years, most of us have moved at least once, so you can imagine the impossibility of trying to connect with everyone again. I've offered Rosie space in future issues of the BR to give a progress report, or other news of the project.

The question of TOTAL POSTAL OVERLOAD does not apply to the above project alone. Rudy Rubberoid of The Rubber Fanzine states, "I am dealing with it by sending out xeroxed form letters for an (undefined) while, ignoring attempts at new contacts and having no pity on the trivial and dull." He then confesses that while this sounds harsh, he is actually a softie, and trys to deal with (answer) it all . . "but I simply can't anymore."

The cost of postage is another factor that is curtailing the efforts and enthusiasm of this dedicated and much loved networker. As he states, "right now, even the price of postage hurts." He initially printed and mailed Vol.4#2 to the contributors. A following payday he printed another 150 copies for subscribers and trades. Obviously he is not receiving enough money from subscriptions to cover printing and postage costs.

In this February issue of the Fanzine, Rudi announces an upcoming change of editorship. He plans two more issues (May & Aug) then the zine goes to Steve Bieler. I would guess a major reason for this change is a result of the aforementioned "Total Postal Overload."

Are we still rallying around that old mailart 'unwritten rule?' . . MAILART & MONEY DON'T MIX? If so, WHY? We all know it costs plenty to mail, and when you get into publication, you have to add not only printing costs, but additional postage for the heavier envelopes.

Hey! Even Jon Held is asking (appologetically for sure) for money for his Mail-Art Bibliography that he's producing with considerable support

continued on back page . . .

POSTSCRIPT

The highlight of my teen years was spent taking the Marin County ferry into San Francisco. Once there, my then-best friend and I would take the bus to Haight Street and walk around, thrift shop, get annoyed that no one sold Diet Coke, and mostly sit on a grassy hill that I never remember the sun setting on. It was hazy and never dull—a wide range of characters from quirky young professionals to crusty iconoclasts who had seen better days passed by becoming fodder for inside jokes that would last years. It was a center for counterculture, though at that time, my idea of what culture it was countering was vague.

I left California in 2011. Some of these memories are tucked away as rare souvenirs of excited teenagehood, but mostly, I've left, moved on, and haven't looked back. Recently, a focus of my work has been revisiting California once again, this time, understanding the nuances of where its reactionary vibe originates. While I've now lived in New York City longer than anywhere else, I've come to appreciate San Francisco in a new light, especially after writing this story, and seeing that the city's public art events and eccentric style have deep roots of identity, self expression, and community. All valuable characteristics that deepen a lived experience.

Logistically, this story has a precursor. The year prior I wrote about contemporary illustrators raising funds for the USPS by sending mail during the 2020 election. It was published on AIGA Eye on Design, a graphic design publication. Through speaking to modern-day mail artists, I researched the movement's origins and learned about Ray Johnson's Correspondence School and the many mail artists connected to it, Anna Banana being a favorite. Her memorable name, vibrant style and consistent visual language immediately grabbed my attention, and I thought she would make a captivating person to profile.

I originally wrote a version of this story as part of my masters project at Columbia University's Journalism School. When it came to developing story ideas for the project, I was interested in combining my background as a web designer and design teacher to the project. I love the way networks can take a lot of forms (analog and digital) and connect internet history to visual culture and art. I hadn't come across many profiles that humanized the formation of such networks and thought it would be an exciting opportunity to profile some of the women behind them, continuing to add more representation to the design canon. Anna Banana immediately came to mind. Contacting her required a little digging—her website was down and email was not publicly listed. Using the Internet Archive's Way Back Machine, I was able to access an earlier iteration of her site that had her email address listed. Amazingly, the email worked and Ms. Banana and I began our correspondence. Through a phone call and email, she told me about Irene Dogmatic, who I later located through her personal site. Irene Dogmatic later connected me with two curators who in turn connected me with Jennie Hinchcliff. Through them all, I've learned a lot about womanhood, connectivity, and how valuable documenting your life and sharing it with others can be. As of June 2023, Anna Banana is 83 years old and living in Canada. Her memory is not as precise as it once was, though her emails are still vibrant, fun, stylized, and enthusiastic. I'm happy that this piece can help remind her of time that has passed and I'm thankful for the time she spent retelling her stories to me in our emails and phone exchanges.

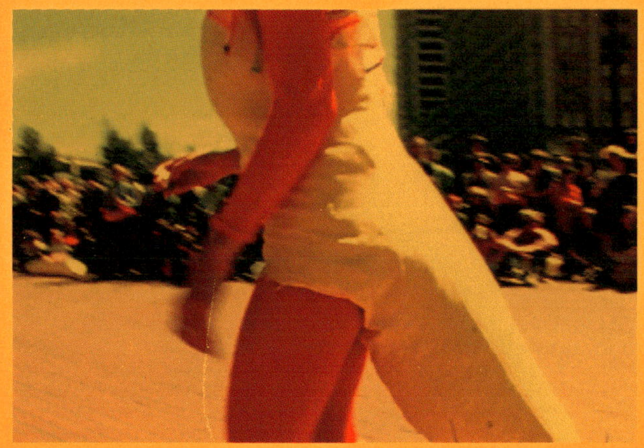

ISBN 979-8-218-22127-0